Cyprustar

The Earth is concave

Summary

Introduction

"Flat Earth" landed on my computer screen close December 2016. I must not be the only one in this case to have had my YouTube home page full of videos selling me that the Earth is a disc covered with a glass dome sailing in the universe.

Today when I write these words I wonder how I could believe for six months to all that. And yet so. Hard like iron, to berate myself with everyone, to want to impose this idea, as if I had done a flight in MIG to be able to support this delirium.

This delusion why? Because all this is science fiction, it's inconceivable and it's a psychological operation extremely strong and powerful. I will even say dangerous for the mind.

In 40 years of life, it can be said that I will have been a "victim" of US government twice;

- In November 1990, in France near the futur Disneyland Paris, I make an observation very close to a massive triangular UFO with my mother and my brother. (Surely an anti-gravity machine of the army American). I am convinced that extraterrestrials exist.
- In December 2016, I am "persuaded" that the Earth is flat.

I can prove that the first point (CIA docs in support) is a psyop, not yet the second because the operation is in progress :).

It is obvious that 'the' flat Earth 'has been pushed heavily on the front of the internet scene in order to unpack some of the population. And it works !

It works so hard that communities, groups create, but we also realize that it divides and feeds hatred. There are also people who will never have the tilt and who will wait until there is an official government announcement to believe in anything again.

So back to the end of 2016 when I open a video probably called to the time; The Earth is flat! Where a guy had to film the horizon and show us something that could not be possible on a convex Earth.

One, two, three, four videos, and we start looking by
oneself and one says oneself; Ok no we can not be on a ball
there's something wrong. "
And what happens is that we just go in one direction;
In the direction of the flat earth. We defend it, we sell it
because we love him. Because we like to think we are hiding the
walls surrounding the Earth and that NASA lied to us about the
trips on the moon.
But it's not exactly that.
July 2017, It's been a month now that Fred Avès and
other globalists send me videos to explain to me that the
stars can not turn in the same direction everywhere on the
planet. Added to that the distances between Antarctic bases
seemed impossible to me, after a while I pick up
totally flat Earth.
One night, I told myself that I had to stop and I closed the
video that I watched.
That's when I realized it was a psychological operation.
When I saw how much there was behind all these productions
on the flat Earth, a huge production. A lot of money
injected into documentaries, with information completely
wrong (Eric Dubay and his nets of stars) but easily verifiable
eventually.
At that time I was deconditioned of the convex globe lost in
the infinite universe but I no longer believed that the earth was
a plane.
If I stayed so long (6 months) on this model, I knew all
the days, despite everything, that it was not the end; that the
Earth does could not have this precise form.
Returning to the desktop of my computer I realized that the
truth was even more mysterious finally. Because even when
we believe in what I believe, we have not finished asking
ourselves questions in permanently.
That's how I came across Lord Steven Christ, the little genius of
the concave earth.
I had to open a presentation video then I chained
with the rectilinear.
It was finally the real revelation.

Curvature tests were explained by the positive curvature
concave, the distance and size of the stars and the paradigm
were logical; That's why the man had never been able to get out
of the Earth. He had not walked on the moon.
Antarctica returned to its original form and stars
were turning well in one direction in the northern hemisphere
and in a other in the south.
I knew I was right.
The reality is there; We live in a closed world, a globe
"Open" at the poles, our space is in the hollow of the Earth, the
stars are very small and very close, our universe is surely
non Euclidean and we do not know what's beyond the poles ...

1) Presentation of the concave earth

The Earth is concave, it means we live inside of the earth.

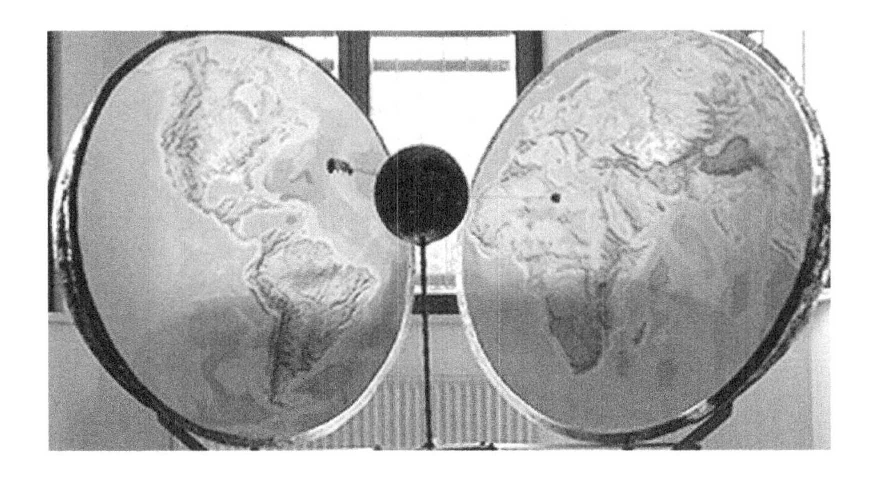

Dimensions

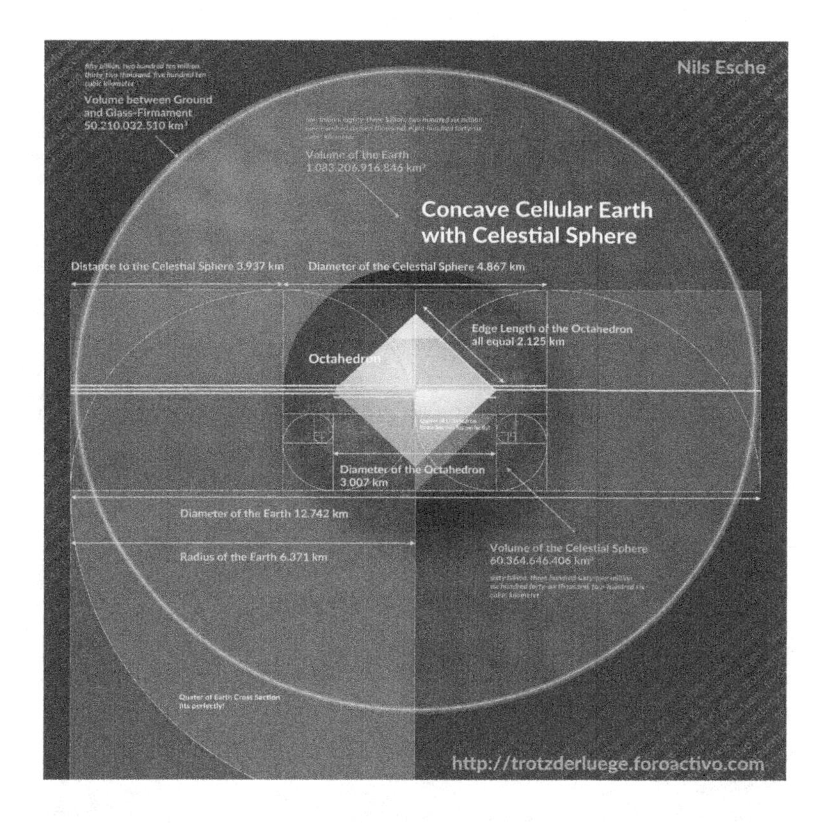
Nils Esche
Volume between Ground and Glass-Firmament 50.210.032.510 km³
Volume of the Earth 1.083.206.916.846 km³
Concave Cellular Earth with Celestial Sphere
Distance to the Celestial Sphere 3.937 km
Diameter of the Celestial Sphere 4.867 km
Edge Length of the Octahedron all equal 2.125 km
Octahedron
Diameter of the Octahedron 3.007 km
Diameter of the Earth 12.742 km
Radius of the Earth 6.371 km
Volume of the Celestial Sphere 60.364.646.406 km³
http://trotzderluege.foroactivo.com

I think that the Earth has a total thickness of about 800 km of depth.

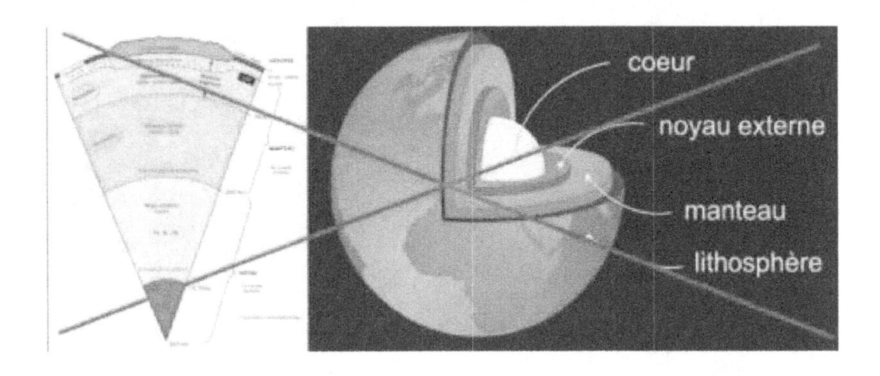

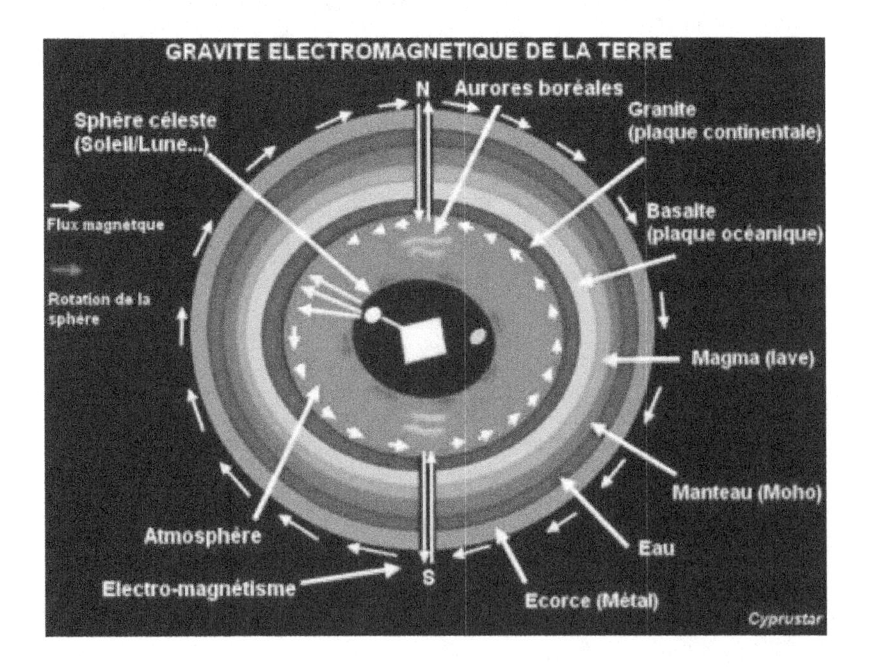

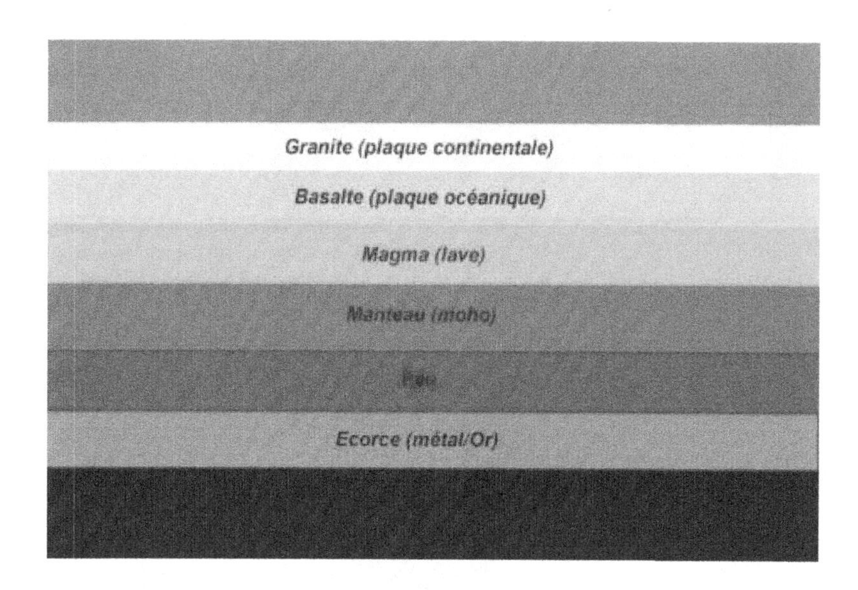

800 km of deep

2) It is the celestial sphere that turns and not Earth

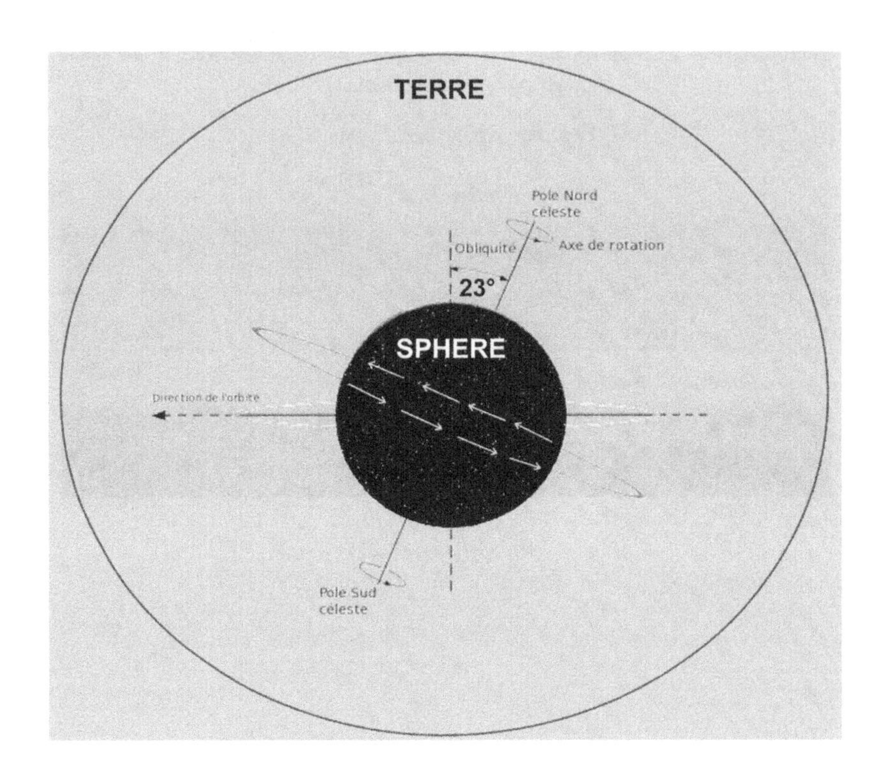

3) The experience of lead wires

Geodetic surveys were carried out at the Tamarack mines, near Calumet, Michigan. The wells were chosen and the lead wires exactly 1295 meters long were hanging in each well. Measurements have begun and experience has shown that the lines, unlike the expectations, were farther apart at the bottom of the wells and less up! The center of gravity is not, as previously believed, in the center of the Earth, but must actually be at above the surface of the Earth; in space (the celestial sphere in reality).
At a depth of 1609 meters, it was found a gap of 20.32 cm between 2 lines.

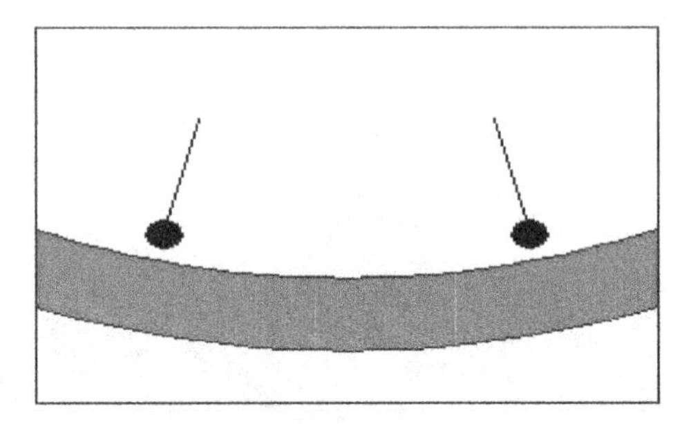

So experience showed that the Earth actually had a curvature concave and non-convex.

4) The rectilinear

MODERN ASTRONOMY OVERTHROWN

The Earth is the great "Womb of Nature." Therefore, life is developed inside. We inhabit the world. This means that we live on the inside surface. The universe is cellular. The discovery that it is cellular was made by **Koresh in 1870**, and his discovery marks the greatest epoch in all science. All the heavenly bodies are within a physical environ 8000 miles in diameter. This completely upsets all modern science and corroborates the truth of the Hebrew and Christian Scriptures. The astronomy of the Bible is true.

Five Sections of the Rectilineator in Adjustment.

Inspired references to the sun, moon, and the earth are correct. The latest edition of **The Cellular Cosmogony** is compiled from the wonderful articles written on the subject by **Koresh**, extending over a number of years, to which have been added the experiments and investigations conducted by the Koreshan Geodetic Staff. Paper cover, $1.00 per copy; cloth, $1.75. Also an interesting Game of Cosmogony, 52 cards in neat case, answering over 200 questions relative to Koreshan Astronomy, sent postpaid for 50 cents.

THE GUIDING STAR PUBLISHING HOUSE
(Department of The Koreshan Unity, Inc.)
Estero, Lee County, Florida

The rectilinear was a geodesic measurement tool developed by geodesist Ulysses Grant Morrow.
The supports of 3m 65 were erected on the beach of 6 km from the Bay of Naples, Florida, southbound, parallel to the littoral.

Original measures

TABLE SHOWING ALTITUDE OF AIR LINE ABOVE DATUM LINE AT EVERY STATION OF SURVEY, WITH THE MEASURES COMPARED WITH THE CALCULATED CURVATURE

Dates of Measurements on Tide Staffs	Distance in Feet From Beginning	Distance in Miles	Number of Adjustments	Number of Tide Staff on Beach	Inches Altitude of Air Line Above Fixed Datum Line	Distance of Air Line Below Secondary Datum Line, inches	Calculated Ratio of Concave Curvature, Inches	Difference Between the Ratios, inches
Mar. 18	0	0	0	1	128	0	0	0
" 19	660	⅛	55	2	127.85	.15	.125	.025
" 23	1,320	¼	110	3	127.74	.26	.5	.24
" 24	1,980	⅜	165	4	126.625	1.375	1.125	.25
" 25	2,640	½	220	5	126.125	1.875	2	.125
" 27	3,300	⅝	275	6	124.125	3.875	3.125	.75
" 30	3,960	¾	330	7	123.675	4.375	4.5	.125
" 31	4,620	⅞	385	8	121.57	6.43	6.125	.305
Apr. 1	5,280	1	440	9	119.98	8.02	8	.02
" 2	5,940	1⅛	495	10	117.875	10.125	10.125	.0
" 8	6,600	1¼	550	11	116.44	11.56	12.5	.94
" 9	7,260	1⅜	605	12	113.69	14.31	15.125	.815
" 13	7,920	1½	660	13	111.07	16.93	18	1.07
" 14	8,580	1⅝	715	14	107.19	20.81	21.125	.315
" 14	9,240	1¾	770	15	104.69	23.31	24.5	1.19
" 15	9,900	1⅞	825	16	101.69	26.31	28.125	1.825
" 16	10,560	2	880	17	97.38	30.62	32	1.38
" 24	11,220	2⅛	935	18	93.44	34.56	36.125	1.565
" 26	11,880	2¼	990	19	85.32	42.68	40.5	2.18
" 27	12,540	2⅜	1,045	20	79.75	48.25	45.125	3.125
May 8	13,200	2½		21	74	54	50	4
" 8	13,860	2⅝		22	68	60	55.125	4.875
" 8	14,520	2¾		23	63	65	60.5	4.5
" 8	15,840	3		24	53	75	72	3
" 8	21,780	4⅛		25	0	128	136.125	8.125
RETURN SURVEY								
" 9	12,540	2⅜	1,084	20	79.75	48.25	45.125	3.125
" 11	11,880	2¼	1,140	19	85.47	42.53	40.5	2.03
" 11	11,220	2⅛	1,194	18	93.68	34.32	36.125	1.805
" 11	10,560	2	1,250	17	97.13	30.87	32	1.13

Measurements converted to the metric system

Test géodésique 1897 – LE RECTILINEATEUR
Ulysses Grant Morrow
Naples, Floride, USA

DATE	DISTANCE EN METRE	HAUTEUR AU DESSUS DE LA REF EN METRE	HAUTEUR SOUS LA 2e REF EN METRE	COURBURE CONCAVE EN METRE
18/03	0	3,25	0	0
19/03	201	3,24	0,301	0,003
23/03	402	3,23	0,661	0,01
24/03	603	3,20	0,024	0,02
25/03	804	3,20	0,025	0,05
27/03	1005	3,14	0,076	0,07
30/03	1207	3,12	0,101	0,11
31/03	1408	3,07	0,152	0,15
01/04	1609	3,02	0,203	0,20
02/04	1810	2,97	0,254	0,25
08/04	2011	2,94	0,279	0,31
09/04	2212	2,87	0,355	0,84
13/04	2414	2,81	0,406	0,45
14/04	2615	2,71	0,508	0,53
14/04 bis	2816	2,64	0,584	0,62
15/04	3017	2,56	0,660	0,71
16/04	3218	2,46	0,762	0,81
24/04	3419	2,36	0,863	0,91
26/04	3621	2,15	1,06	1,02
27/04	3822	2,00	1,21	1,14
08/05	4023	1,87	1,37	1,27
08/05 bis	4224	1,72	1,52	1,40

<u>CONCLUSION</u>: Pour 4224 mètres, il y'a 1,40 mètre de courbure concave.

When we compare 4224 m on Earth curve calculator, we falls exactly over 1.40 meters.

5) The curvature of the Earth

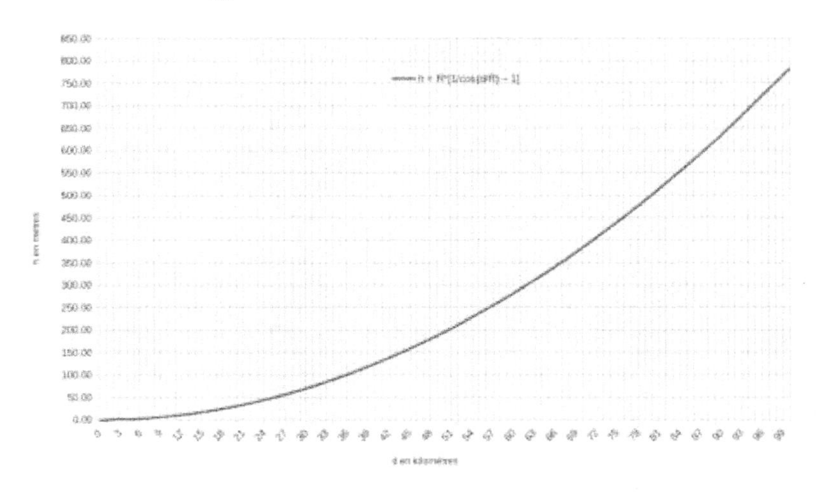

Distance in meters Concave curvature in meter

50 m □ □ □ □ □ □ □ □ 0.0002 m (0.2 mm)
100 m □ □ □ □ □ □ □ □ 0.0008 m (0.8 mm)
500 m □ □ □ □ □ □ □ □ 0.019 m (19 mm)
1000 m □ □ □ □ □ □ □ □ 0.078 m (78 mm)
2000 m □ □ □ □ □ □ □ □ 0.31 m (31 cm)
3000 m □ □ □ □ □ □ □ □ 0.70 m (70 cm)
4000 m □ □ □ □ □ □ □ □ 1.25 m
5000 m □ □ □ □ □ □ □ □ 1.96 m
6000 m □ □ □ □ □ □ □ □ 2.82 m
7000 m □ □ □ □ □ □ □ □ 3.84 m
8000 m □ □ □ □ □ □ □ □ 5.02 m
9000 m □ □ □ □ □ □ □ □ 6.35 m
10,000 m □ □ □ □□ □ □ 7.84 m
20,000 m □ □ □ □□ □ □ 31.39 m
30,000 m □ □ □ □ □ □ □ 70.63 m
40,000 m □ □ □ □□ □ □ 125.56 m
50,000 m (50 km) □ □ □ □ 196.19 m
100,000 m (100 km) □ □ □ 784.75 m

It is natural to understand why today still many people think that the Earth is flat. It can be explained by the very negligible curvature. Indeed, a man average measurement 1.70 m. Earth, it bends so positive of 1.25 m every 4000 meters! So we do not surrender account that we live in a globe.

6) Unable observations on a world

Le mont Canigou

Infrared proves that Mount Canigou is not a refraction.
A slew of photos and videos prove that if we see the Mount Canigou is that he is really present.
In particular this photo taken in infrared by Bruno Carrias.
This picture tells us everything: It was taken from the bars rocks of the Holy Spirit at 384 m of altitude to 265 km of the Mount Canigou (2784 m) on the 7/03/2018.

This photo is proof that it is not a refraction that one sees but a physical relief.
Infrared proves it.
At an altitude of 384 m, an object targeted at 265 km is hidden by 2985.04 m of curvature!

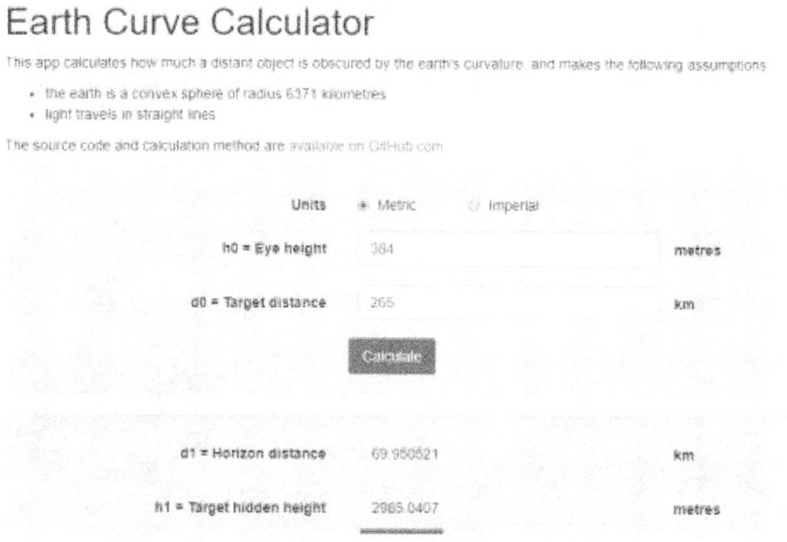

Earth Curve Calculator

This app calculates how much a distant object is obscured by the earth's curvature, and makes the following assumptions

- the earth is a convex sphere of radius 6371 kilometres
- light travels in straight lines

The source code and calculation method are available on GitHub.com

Units	Metric / Imperial	
h0 = Eye height	384	metres
d0 = Target distance	265	km
	Calculate	
d1 = Horizon distance	69.950521	km
h1 = Target hidden height	2985.0407	metres

This is proof that we live in a globe and not on it,
because the Mount should be totally obscured by the curvature
either saying convex of the Earth.

The Barre des Ecrins seen at 440 km

The Barre of the Ecrins (4.102 m) in the middle of the image
Photo taken from Pic de Finestrelles (2,820 m) at 440 km,
before sunset.
According to the calculation, the convex curvature should hide
4919 meters.
So we should not be able to see the bar of the boxes
at 4102 meters! The curvature is concave.

The picture of the US army

The photo is from "FotoMagazin" n ° 11/1954.
Dr. Fritz Neugass comments:
"A new telephoto lens of the US Army"
The Optics Research Division of the Signal Corps of the US military has just created a new camera, specially designed to take pictures at a distance of 50 km.
The lens has a focal length of 254 cm, is 1 m long and has a diameter of 24.13 cm, it has been corrected for the use of a infrared film.
Using this lens, it is quite easy to analyze the terrain to a distance of 10 to 20 km and to distinguish between weapons, fortifications and transport.
The disadvantage of such a telephoto lens is the complete elimination of the perspective.
The photo reproduced shows the Empire State Building and the outline of Manhattan at a distance of 41.8 km
At the bottom of the Empire State Building, a large hotel is visible on Coney Island is 20.9 km from the camera.

Sandy Hook Lighthouse, in the foreground of the photo, is only 6.4 km from the camera.
The photo allows a view of the rooftops of the port city behind she, optically understands the Brooklyn Peninsula and another sea inlet and clearly shows the skyscrapers of Manhattan.
If this Earth were a convex ball, all this should be 100 m under the horizon.

The picture taken on the teleobjective

7) The true solar system (New astral triangulation method)

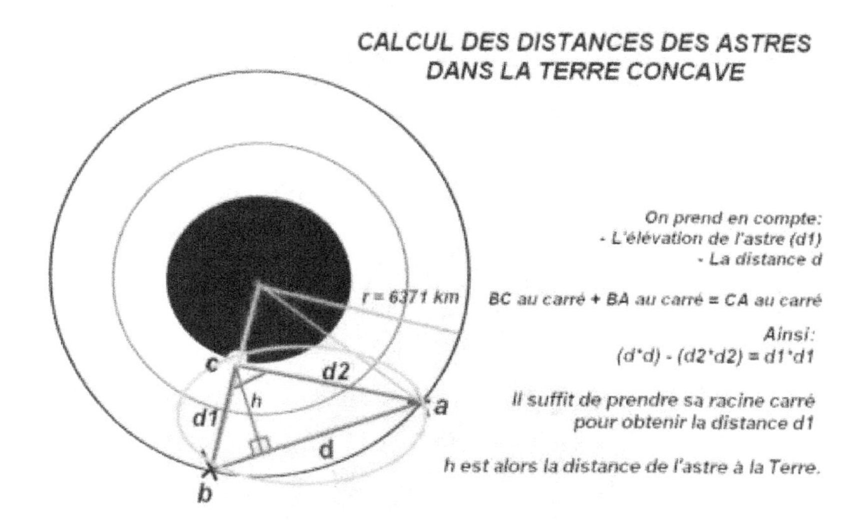

A method of triangulation finally allowed me to find the approximate distance of all the stars of the celestial sphere. She to take two points on Earth:

The first point sees the star rising, the second in b sees it at its solar noon, meaning that it is facing south.

Ma méthode consiste à systématiquement calculer avec la formule Vincenty et le théorème de Pythagore, la distance d en l'additionnant à l'élévation observée au point b.

(d au carré*d au carré) + (d1 au carré*d1 au carré) = d2 au carré

Il suffit de prendre la racine carré pour obtenir d1. On obtient ainsi la hauteur de l'astre h.

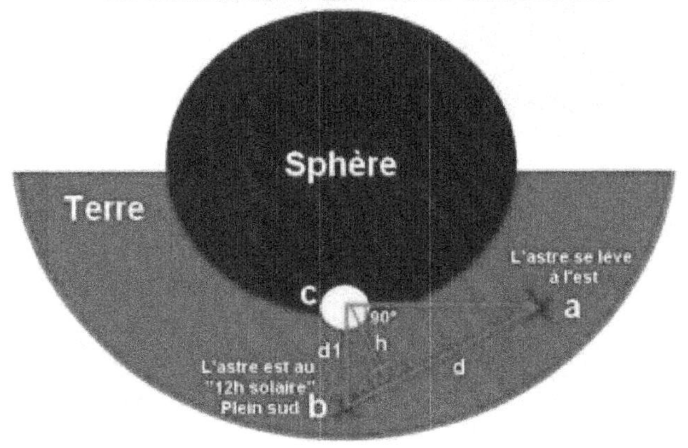

Thus the right angle is systematically formed. Must therefore calculate the distance between these points. This is the theorem of Pythagoras.
To this we add the magnitude of the star at solar noon of point b.
We then obtain the height h of the star thanks to the magnitude of the star taken on Stellarium. This software also allowed me to make the precise points of observation.

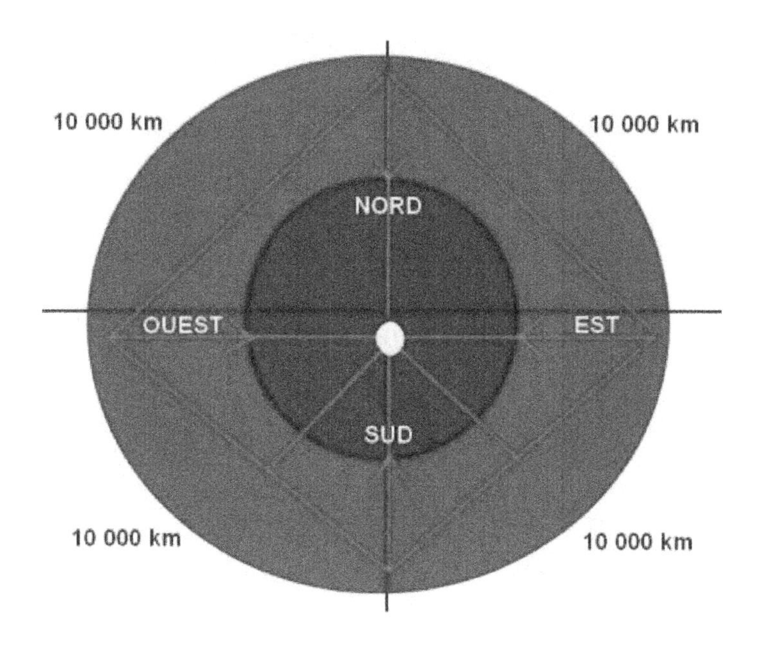
10 000 km
10 000 km
NORD
OUEST
EST
SUD
10 000 km
10 000 km

Triangulation

March 2, 2018

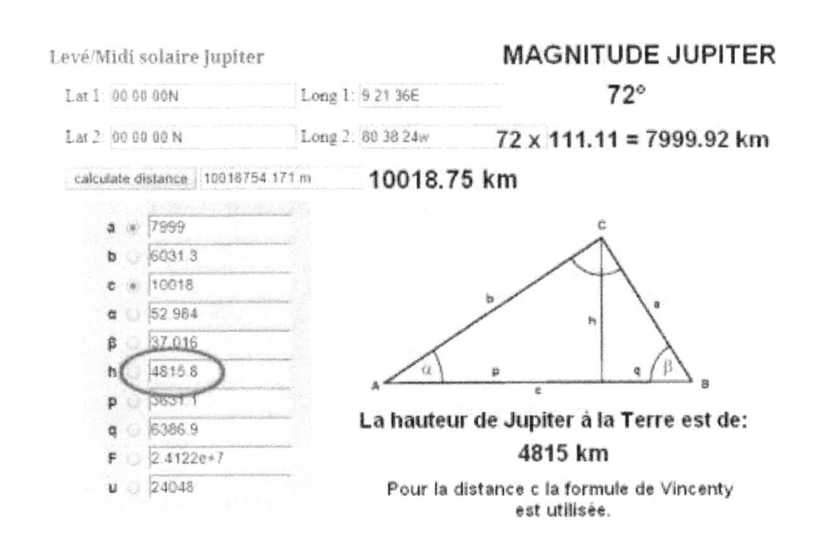

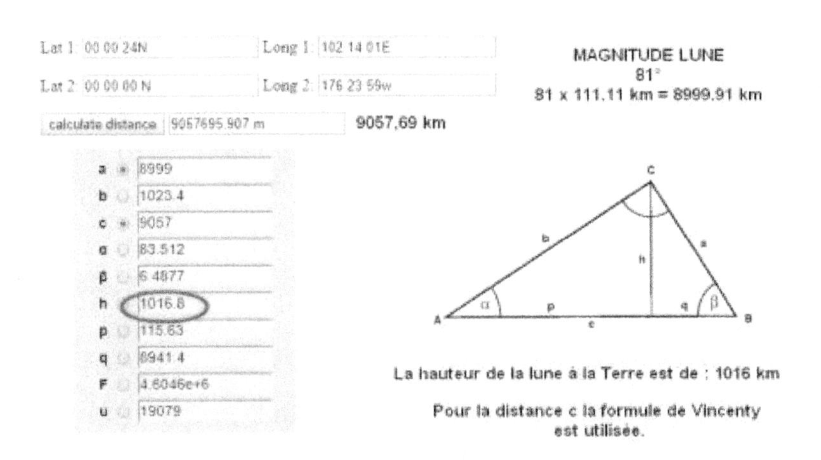

Lat 1: 00 07 00S | Long 1: 09 21 36E

Lat 2: 0 39 42 N | Long 2: 84 14 23W

calculate distance | 10419692.530 m

10419,59 km

a	7333.3
b	7402.1
c	10420
α	44.732
β	45.268
h	5209.6
p	5258.5
q	5161.1
F	2.7141e+7
u	25155

MAGNITUDE MARS

66° = 111.11 x 66 = 7333,26 km

Pour la distance c la formule de Vincenty est utilisée.

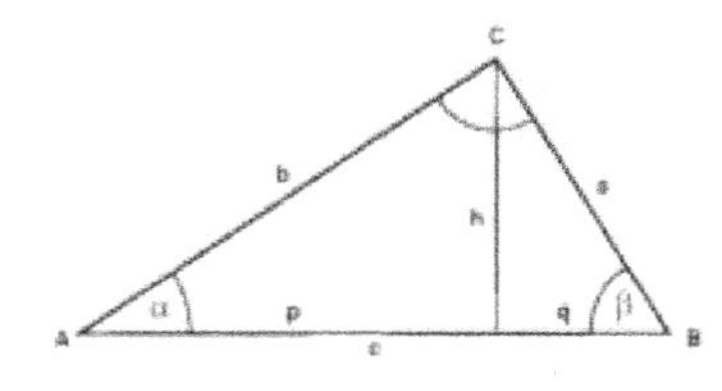

Hauteur de Mars à la Terre : 5209 km

Lat 1: 00 39 42S | Long 1: 05 02 24E

Lat 2: 00 00 00 N | Long 2: 102 14 23E

calculate distance | 10820167.953 m

10820,16 KM

a	9111
b	5836.5
c	10820
α	57.356
β	32.644
h	4914.6
p	3148.3
q	7671.9
F	2.6588e+7
u	25768

MAGNITUDE MERCURE
82° =
82 X 111.11 = 9111.02

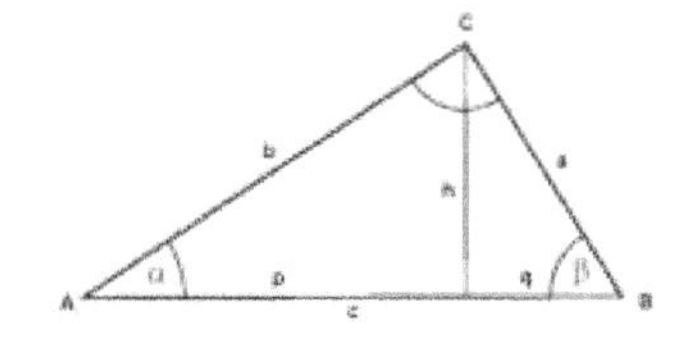

La distance à la Terre de Mercure est de: 4914 km.

Pour la distance c la formule de Vincenty est utilisée.

Lat 1	00 00 00N	Long 1	09 21 36E
Lat 2	0 39 42 N	Long 2	80 38 24W
calculate distance	10018751.947 m	10018 KM	

a	7666.6
b	6448.5
c	10018
α	49.932
β	40.068
h	4936
p	4150.9
q	5867.1
F	2.4719e+7
u	24133

MAGNITUDE SATURNE=69°
69° x 111,11 km = 7666,59 km

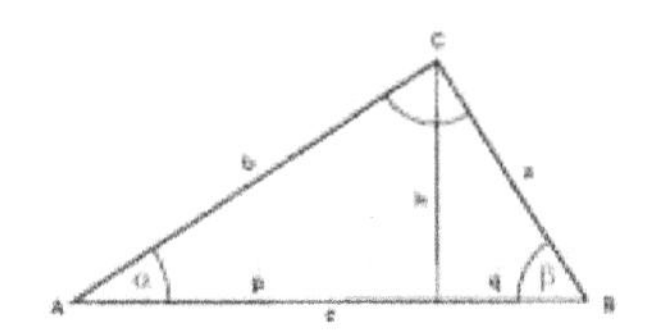

Pour la distance c la formule de Vincenty est utilisée.

Hauteur de Saturne à la Terre : 4935 km

Lat 1:	00 00 24N	Long 1:	176 23 59W
Lat 2:	00 00 00 N	Long 2:	80 38 24w
calculate distance	10659923.512 m	10659,92 km	

a	8888
b	5885.4
c	10660
α	56.489
β	33.511
h	4907.1
p	3249.5
q	7410.6
F	2.6154e+7
u	25433

Magnitude soleil : 80°

80 x 111.11 = 8888.8 km

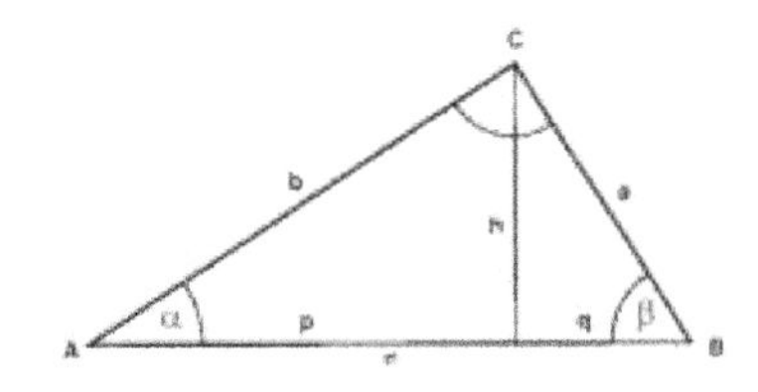

Hauteur du soleil à la Terre : 4907 km

Pour la distance c la formule de Vincenty est utilisée.

Lat 1: 00 00 00N Long 1: 9 21 36E

Lat 2: 00 39 42 N Long 2: 102 57 36E

calculate distance 10419475.525 m ***10419,47 km***

Magnitude d'Uranus : 80°
80 x 111.11 km = 8888.88 km

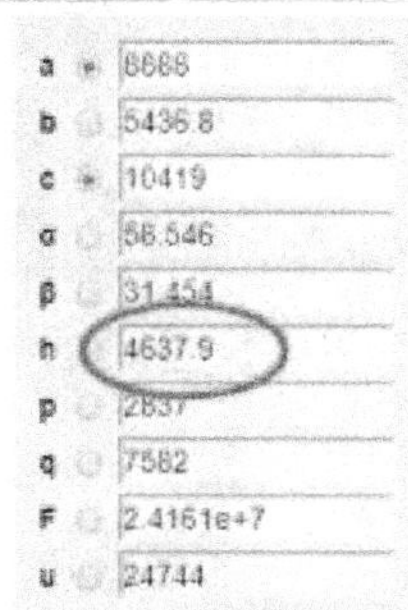

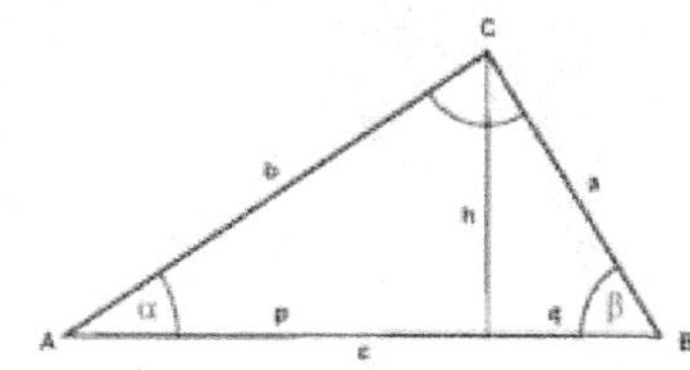

La hauteur d'Uranus à la Terre est de: 4637 km

Pour la distance c la formule de Vincenty
est utilisée.

Lat 1: 00 00 00N Long 1: 09 21 36E

Lat 2: 00 00 00 N Long 2: 102 14 23E

calculate distance 10339323.383 m 10339,32 KM

MAGNITUDE VENUS
84°
84° x 111.11 = 9333.24 km

a	9333.3
b	4448.7
c	10339
α	64.516
β	[illegible]
h	4015.9
p	1914.2
q	8425.2
F	2.0761e+7
u	24121

Pour la distance c la formule de Vincenty
est utilisee.

C, b, a, h, α, p, q, β, A, c, B

La hauteur de Vénus à la Terre
est de 4015 km.

A geo heliocentric system

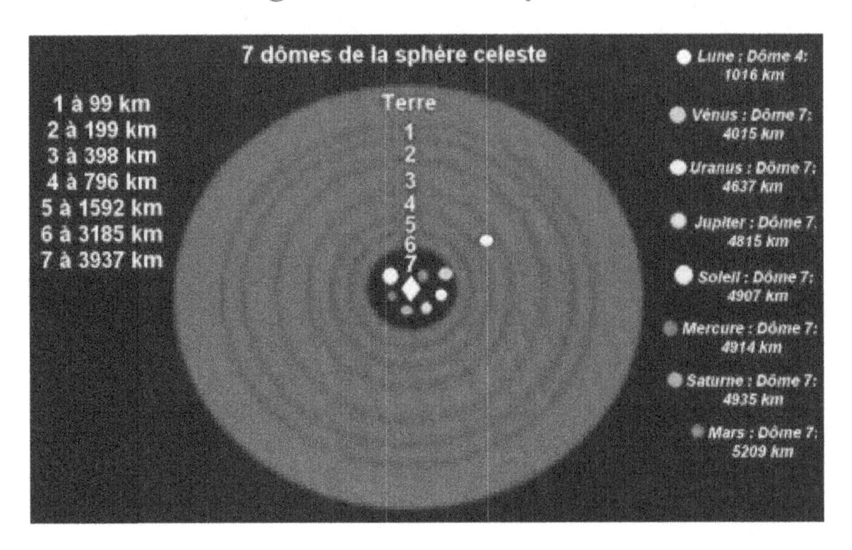

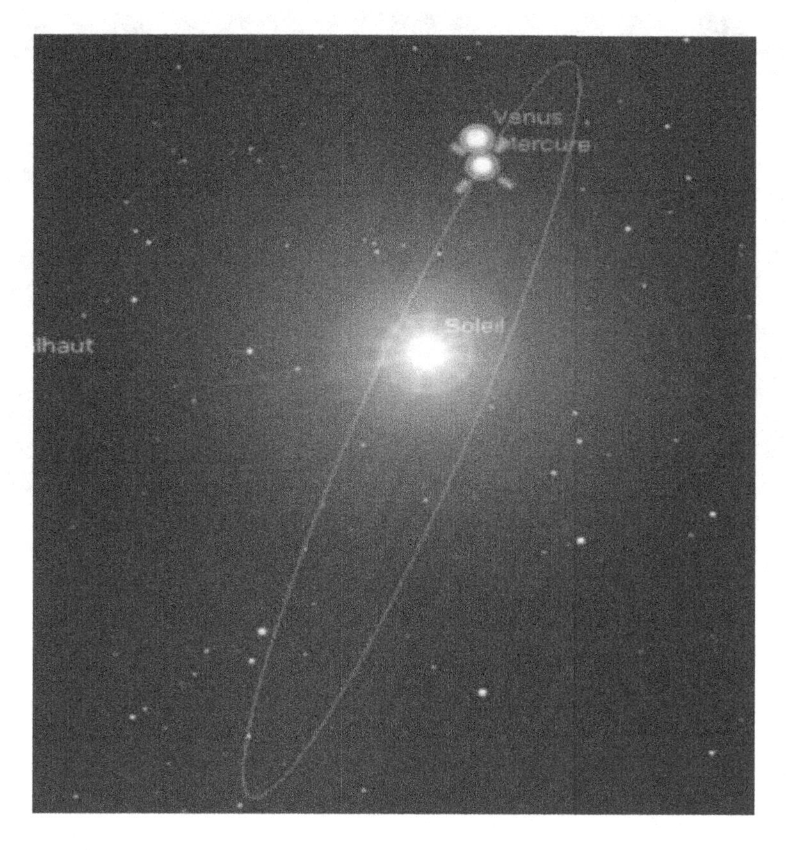

Mercury revolves around the sun

But the sun is turning around the Earth. It has six orbits.

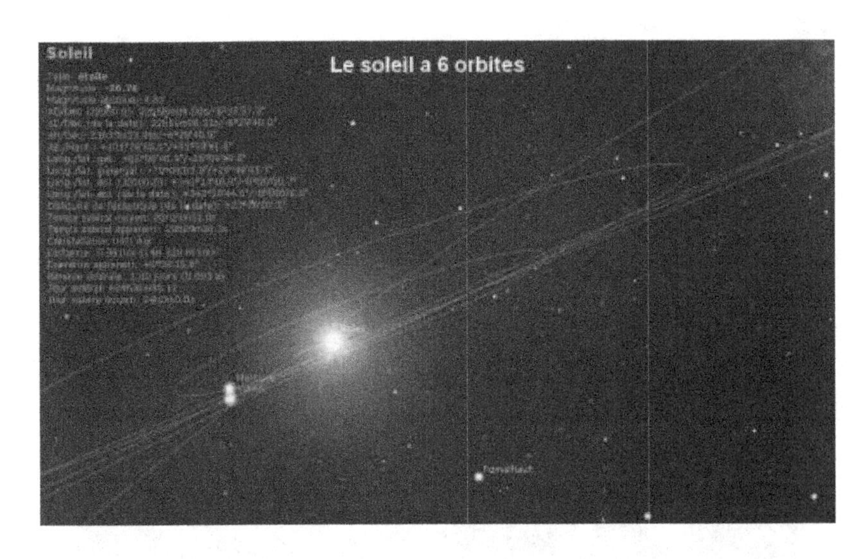

Venus evolves around the sun

Uranus
Venus
Mercure
Soleil

Diameters of the planets

Since we have the apparent diameter of the planets and now their true distances we can get their true diameters.

Diamètres apparents (en minutes et secondes d'arc) du Soleil, de la Lune, des planètes et de planètes naines du Système solaire, observés depuis la Terre

Objet	Minimum	Maximum	Moyen en conjonction inférieure	Moyen en opposition	Réf.
Soleil	31' 27"	32' 32"			1
Mercure	0' 4,5"	0' 13"	0' 11"		2
Vénus	0' 9,7"	1' 6"	1' 0,2"		3
Mars	0' 3,5"	0' 25,1"		0' 17,9"	4
Lune				31' 36"	5
Jupiter	0' 29,8"	0' 50,1"		0' 46,9"	6
Saturne	0' 14,5"	0' 20,1"		0' 19,5"	7
Uranus	0' 3,3"	0' 4,1"		0' 3,9"	8
Neptune	0' 2,2"	0' 2,4"		0' 2,3"	9
Pluton	0' 0,06"	0' 0,11"		0' 0,08"	10

The sun is 46.13 km in diameter

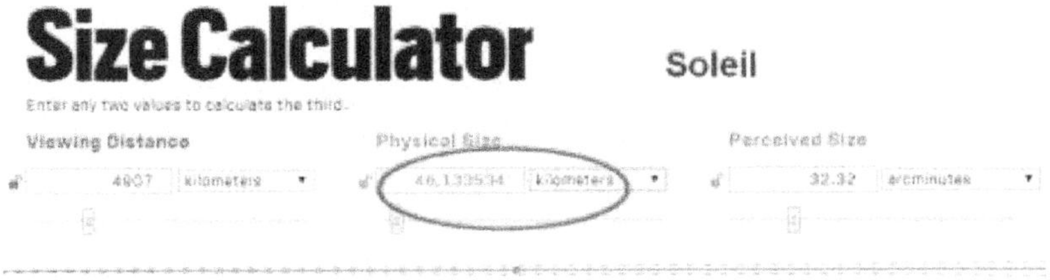

Size Calculator Soleil

Enter any two values to calculate the third.

Viewing Distance: 4907 kilometers

Physical Size: 46.133534 kilometers

Perceived Size: 32.32 arcminutes

Mercury is 185.82 m in diameter

Size Calculator

Mercure

Enter any two values to calculate the third

Viewing Distance

Physical Size

Perceived Size

Venus is 1.86 km in diameter

Mars has for diameter 378.80 m

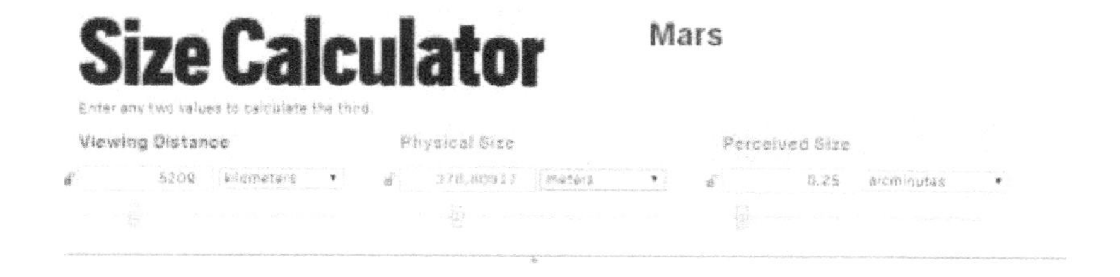

The moon is 9.26 km in diameter

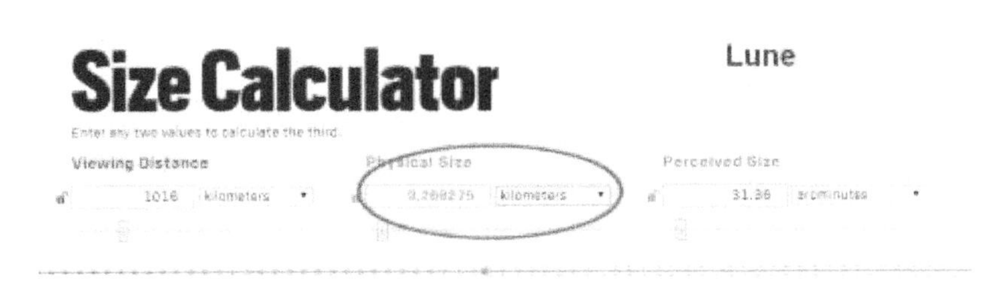

Jupiter: 700.31 m in diameter

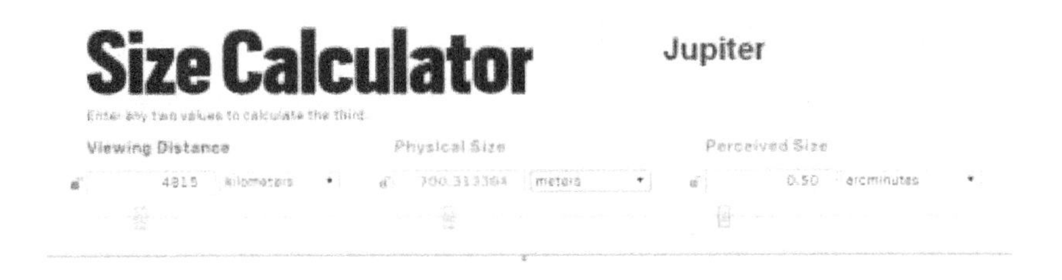

Saturn is 287.10 m in diameter

Uranus is 539.53 m in diameter

Size Calculator

Enter any two values to calculate the third.

Viewing Distance		Physical Size		Perceived Size	
[illegible]	kilometers	539.53846	meters	0.4	arcminutes

8) The seasons

SUMMER SOLSTICE

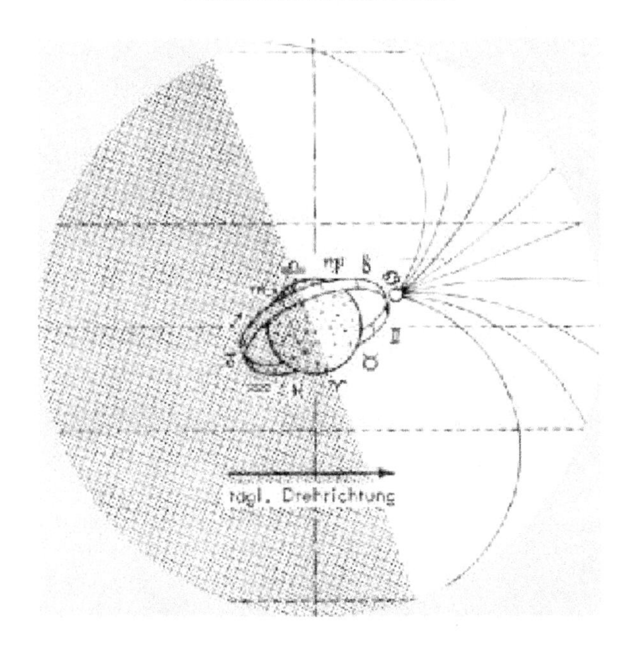

June 21, the sun illuminates the Arctic h24, 7 days out of 7 for 3 month. He is on the tropics of cancer.

SPRING EQUINOXE (MARCH 23) OR OF AUTUMN (SEPTEMBER 23)

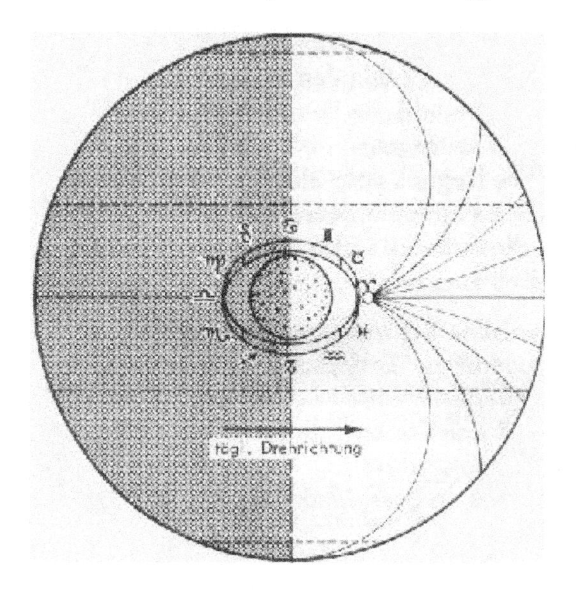

Le soleil is on the equator.

SOLSTICE OF HIVER

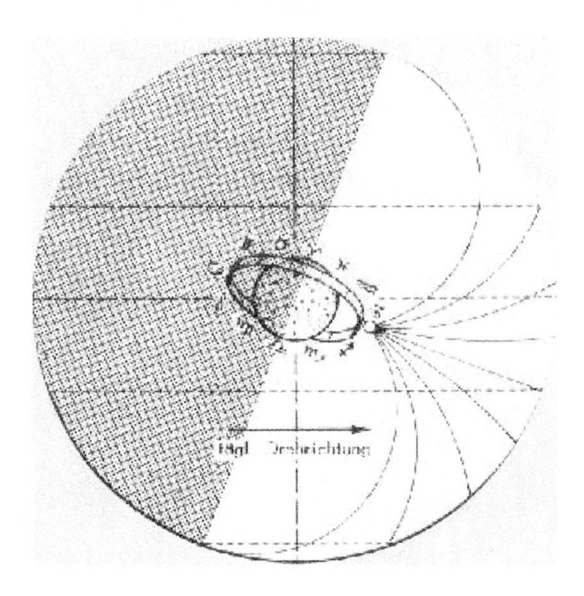

December 21, Le Soleil éclaire l'Antarctique h24, 7 jours sur 7 hanging 3 mois. Il was found in the tropique du Capricorne.

9) THE PHASES OF THE MOON

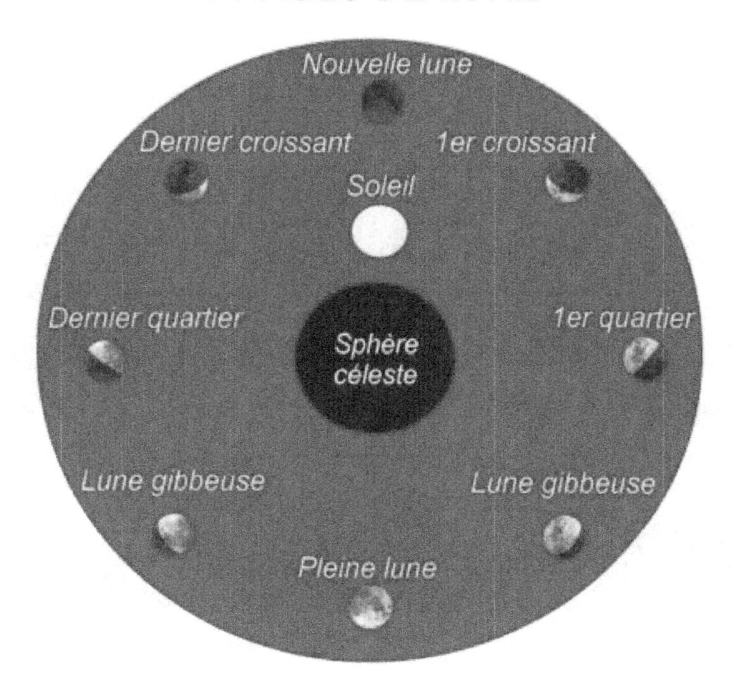

The moon turns inside the Earth and around the sun at 1840 km / h.
The sun, turn at 1612 km / h in average.
So the moon goes beyond the sun creating phases
moon:
New moon: The moon turns its back on the sun and can not not be enlightened.
1st crescent: The moon begins to move away from the sun and this
do this is an embedded foot.
1st quarter: The calm moon made you think about it

Sun
Gibbous Moon: Do not worry, I'll be back soon
nonclassified party.
Full moon: The moon is on the sun and receives all light.

10) Lunaire eclipses are caused by Saturn

A study of lunar eclipses on Stellarium allowed me to understand that the Earth could not be the cause.
Indeed since the celestial sphere is in the hollow of the Earth, it is impossible that it is she who shadows the light that the moon receives.
So I had the idea to make collages of screen prints of Stellarium in order to connect the stars and planets together in order to reach the sun and thus draw a line between him and the moon during an eclipse. From the first collage of photos I noticed that Saturn was on the moon / sun plot. he
turns out that it is absolutely on all lines plotted with each studied eclipse. I chose 4 dates to prove that it is Saturn and not the Earth that causes eclipses lunar: March 4, 2007, July 28, 2018, January 21, 2019, andnJuly 16, 2019.

March 4, 2007

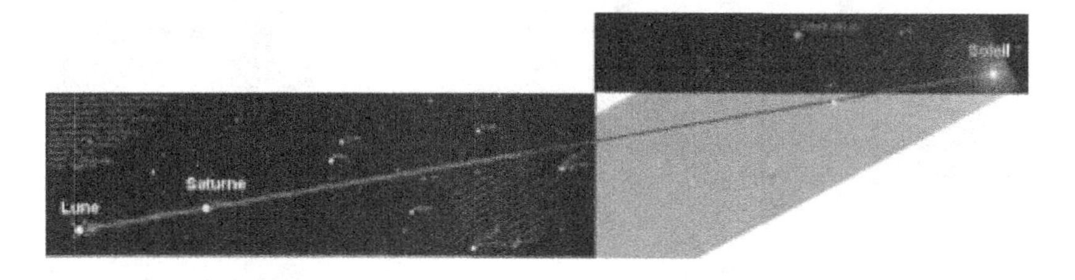

Start: 7:22 am - End: 10:47 am - Duration: 3:25
Positive position of hours: N 83 ° - E 146 °
Saturn, on the moon / sun path.

July 27/28, 2018

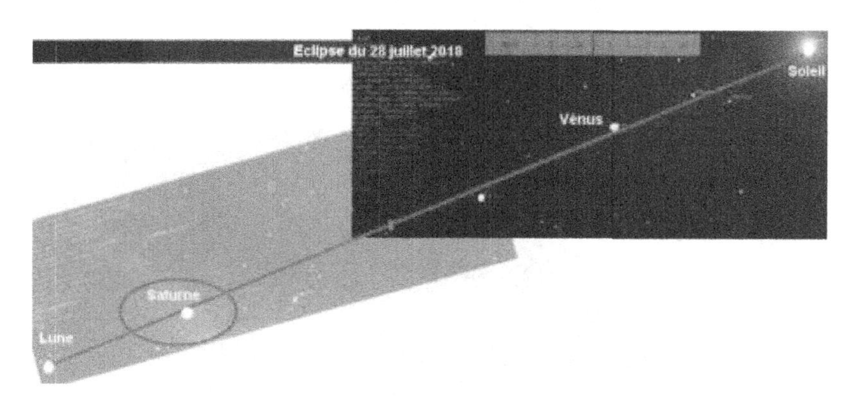

Start: 2h07 - End: 6h10 - Duration: 4h03
Positive position of hours: N 82 ° - E 115 °
Saturn is near the moon, it darkens it. It can not be Venus we will see why further.

January 21, 2019

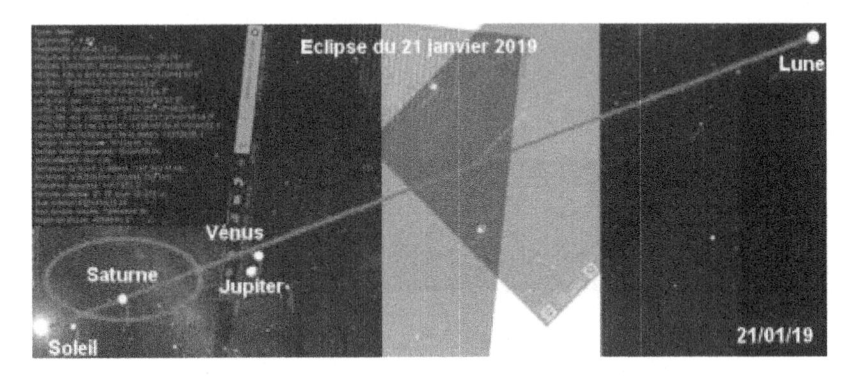

Start: 11h22 - End: 14h38 - Duration: 3h16
Positive position of hours: N 82 ° - E 115 °
This time Saturn is rather close to the sun; it is a barrier to light attributed to the moon.

July 16, 2019

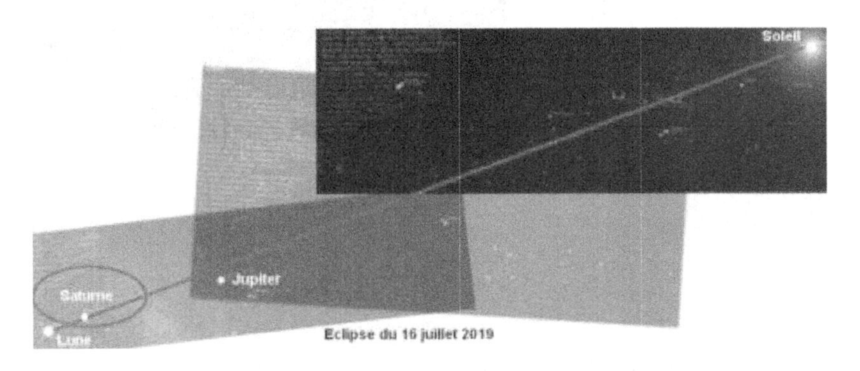

Start: 14h40 - End: 17h40 - Duration: 3h
Position reading hours: S 11 ° - O 81 °
Venus can not be the cause of the lunar eclipse, here's why below.

She is behind the sun from the point of view of the trajectory Moon / Sun.

CONCLUSION

So it seems that this is Saturn the head of eclipses
lunar and not the Earth that contains the celestial sphere. If of Earth we do not see the moon and the sun at the same time it's because
that they are both at the ends of the celestial sphere.

11) Travel speed of the sun and the moon

By observing the zenith of the moon and the sun in two places different, we can calculate the travel time of two stars.

So I took two points on the planet on 1/08/2018 on Stellarium:
10,156 km separate the 2 points.
The moon is at its zenith in Bafia at 4:29 and Balzar in 10:58 p.m..
In 5:31 it travels 10,156 km = 1840 km / h.
The sun is at its zenith in Bafia at 13:32 and Balzar in 7:14.
In 6:18 he travels 10,156 km = 1612 km / h

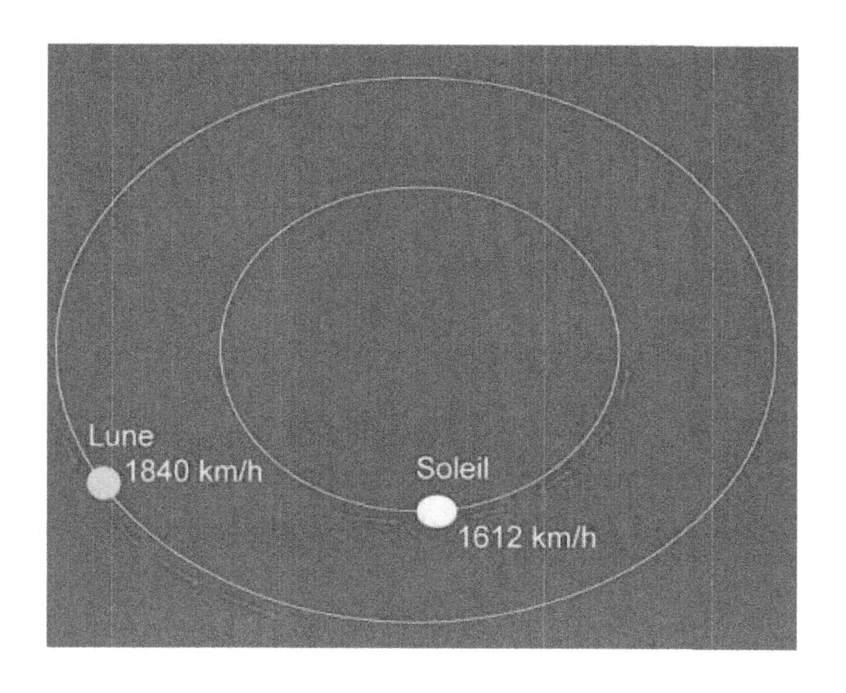
Lune
1840 km/h
Soleil
1612 km/h

12) What is space made of?

There is no infinite emptiness. In our reality, it does not exist. same if you fall in a ravine you will still land on something: The Earth.

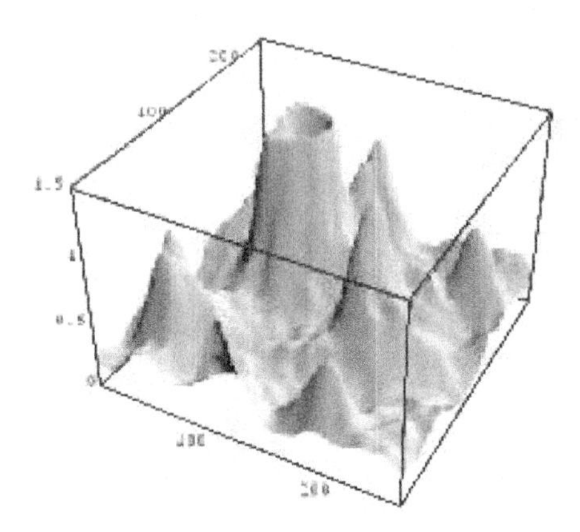

As Aristotle said; "Nature hates emptiness".
Our atmosphere and our space are at the very bottom of our planet.
The established science does not teach us of course that but it informs us about a specific point: In the Universe the temperature reaches -272 ° C. The lowest temperature theoretically possible is the absolute zero at -273.15 ° C.
So what can our heavenly sphere consist of?
And a lot of helium.
Physicists have found that below the temperature critical of 2.17 Kelvin, (ie -270.98 ° C), which is called the point lambda (λ), helium 4 underwent a phase transition.
At this point, helium transforms from a gas into a fluid of zero viscosity, which allows it to flow without loss of energy kinetic. It means that if you put a little helium superfluid in a cup and you spin it, this Helium will literally run forever.
This state of matter is so strange, it has the ability to flow "Up" against gravity, and climb on the sides of a dish.

In the plasma state, helium electrons are not related to core, which leads to a very high electrical conductivity, even when the ionization is partial. The charged particles are very sensitive to electric and magnetic fields.

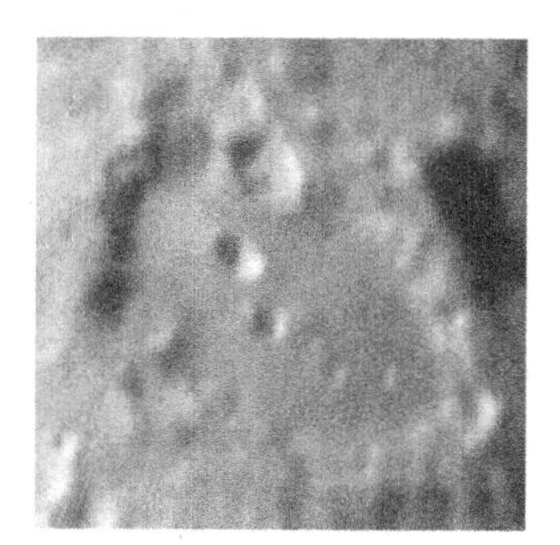

Superfluid helium could also explain the phenomena of lunar waves and transient lunar phenomenon PLT).
So the properties of helium and its ability to become a superfluid at such a temperature and that which we found in space show that the latter has a 90% chance of being consisting of this chemical element.

13) Where do the meteorites come from?

To understand meteorites one must know their compositions as well as that of the sun.

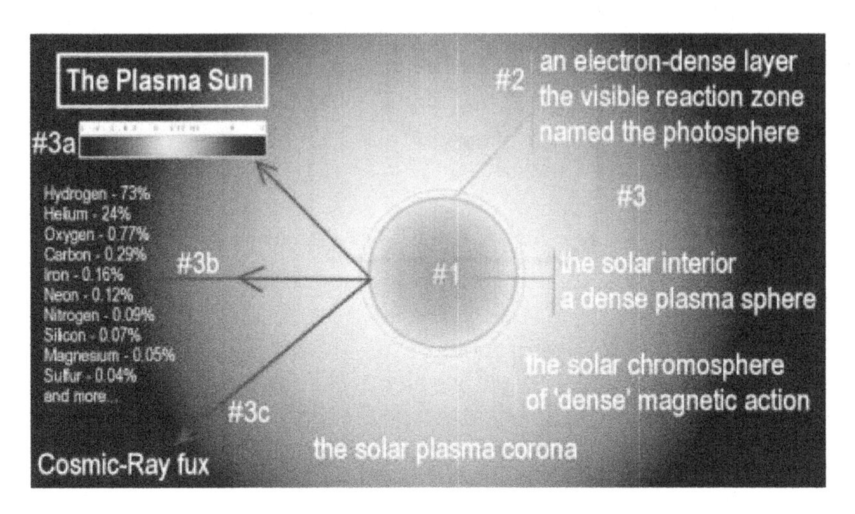

Classification of meteorites

- Chondrites:
Type L (40% of ordinary chondrites) consists of
(Mg, Fe) SiO_3 (hypersthene), Mg_2SiO_4 with up to 25%
Fe_2SiO_4 (olivine), FeS (Troilite), and up to 410%
of the total FeNi meteorite content.
- LL type (10% ordinary chondrites) consists of
(Mg, Fe) SiO_3 (hypersthene), Mg_2SiO_4 with up to 32%
Fe_2SiO_4 (olivine), FeS (Troilite) and up to 3% of meteorite
total FeNi alloy content.
- Type H (40% of ordinary chondrites) is composed of
$MgSiO_3$ with up to 12% $FeSiO_3$, Mg_2SiO_4 with up to
20% Fe_2SiO_4 (olivine), FeS (Troilite) and up to 19% of the
total meteorite FeNi alloy content.
- Carbonates (group C)
- A enstatite (group E); Type E which only represent 2%
of all the meteorites found in the world, are rich in
$MgSio_3$ (enstatite), but most of the iron from this

meteorite is in the form of a fernickel alloy
and FeS (trollite). The rest of the metal is bound to the dioxide component
of silicon with minor amounts of schreibersite (FeNi) and
of graphite.
- other groups (groups K and R);
- differentiated meteorites:
- Achondrites:
- howardites (HOW); Basalt
- eucrites (EUC); Basalt
- diogenites (DIO); Composed of orthopyroxene with 23% orthoferrosilite and 1.5% wollastonite (silica)
- shergottites (SHE); basalts and lherzolitics;
Olivine / aluminum / garnet
- nakhlites (NAK);
- chassignites (CHA);
- angrites (ANG); Olivine, augite and anorthite
- aubrites (AUB); Silicate
- ureilites (URE); Achondrite, olivine, pigeonite
- Iron meteorites, of which there are two types of classification:
- octahedrites; Two alloys of iron and nickel:
relatively low nickel kamacite and richer taenite
- hexaedrites; Iron / nickel
- ataxites; Nickel
- lithosiderites:
Pallasites (PAL); Iron / nickel
- mesosiderites (MES)
- Siderolite; Half of a fernickel alloy and half
of silicates. This is actually a stony meteorite to
metal inlays.

Where do the meteorites come from?

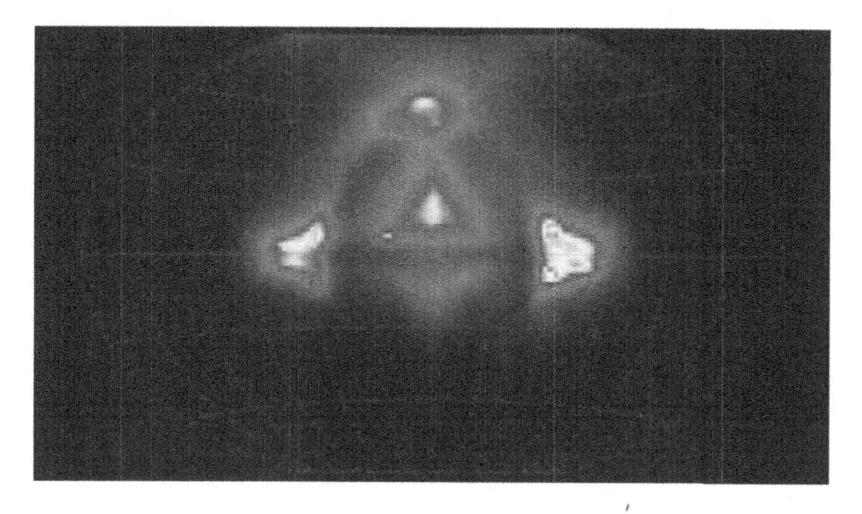

The octahedron (in the middle) highlighting the presence of meteorites (debris)

Meteors come from the Sun and its generator (octahedron / Tokamak), which consist of an alloy of iron, silica, carbon (like chondrites), and hydrogen. the same as a nuclear fusion plant produces debris, the octahedron / sun produces some as well.
Indeed in a tokamak intended to produce energy, the microdebris management, dust and erosion particles produced by disruptions causes maintenance problems, because they can interact with the plasma. So you have to evacuate waste to decontaminate and treat them in accordance with regulation.
So meteorites are waste from the melting of the sun that cross the glass sky (the silica dome) to end expelled to Earth.

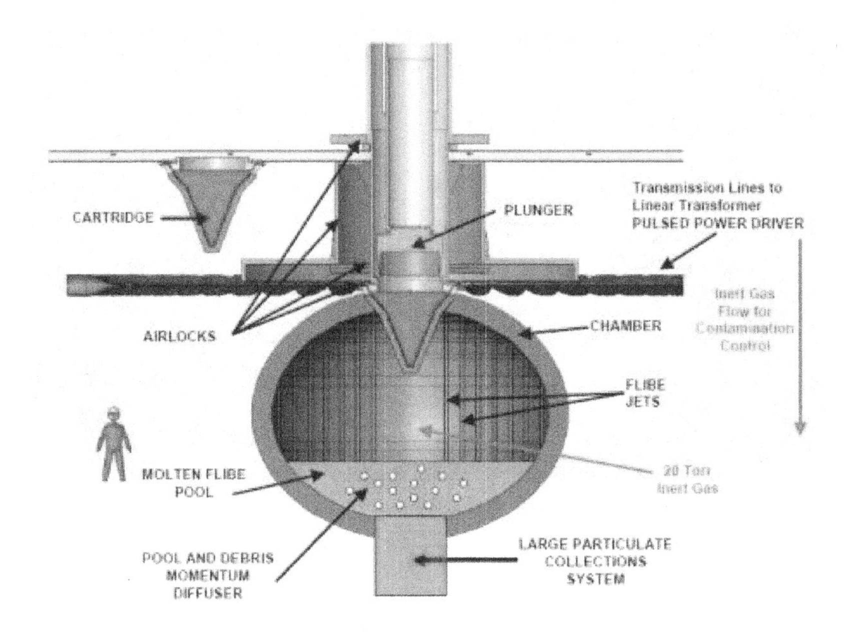

Sectional view of a ZIFE project reactor
What proves that meteorites come from the sun is also this meteorite fell in September 1969 near the small town of Murchison in Australia. The first analyzes revealed that it contained amino acids, cyclic molecules nitrogen, oxygen, carbon and hydrogen, like the others so-called carbonaceous meteorites.

Schreibersite

The Schreibersite is in the form of a plate and is therefore probably plated with fernickelcobalt alloy common already mentioned.
In addition, as graphite and troilite are present in the most meteorites (of iron) and that they come under the form of nodules in fernickel alloy, all meteors have to start life in the form of sulphide expulsions containing graphite which then strikes the plated layer of schreibersite (FeNi) 3P of fernickelcobalt main alloy which constitutes a component of the Sun. Sulfur has a point of low melting and reacts so quickly with the iron of the alloy forming iron sulphide (troilite). This mass of graphite and sulfur to very high temperature (5500 ° C) melts the layer of schreibersite and then melts in the main alloy fernickelcobalt located behind. This melt is then ejected from Sun in a way probably until reaching the layer about 100 km high. At this point, the size of the meteor and / or the angle with which it hits the glass determines how much glass the iron meteor captures and how the glass gets mixture with the meteor, whether by a very total

mixture rare (carbon meteorites) mixture (ordinary chondrites), rare without mixing (ironmelon meteorites) or rare iron meteorites containing no glass, even under chondra form.
The dust collected in the tokamaks proves that meteorites are debris from the sun!

14) Electro-statico-magnetic gravity

The Earth is concave and has in its center the celestial sphere in rotation with the sun and the moon. Our system is a biosphere electric and if the oceans, humans, everything can perfectly to hold on the Earth, it is thanks to the electro-staticomagnetic gravity.

$$\vec{F} = q\vec{E} + q\vec{v} \times \vec{B}$$

Force électrique — Force magnétique

The field exerts on the matter a mechanical action, the force of Lorentz, which is the classic description of the interaction electromagnetic.

There is also:

$$\overrightarrow{Fe} = \frac{q_1 q_2 \overrightarrow{Q_1 Q_2}}{4\Pi \epsilon_0 {Q_1 Q_2}^3} \text{ et } \overrightarrow{Fm} = G\frac{m_1 m_2 \overrightarrow{Q_1 Q_2}}{{Q_1 Q_2}^3}$$

Electrostatic and gravitational forces

Anything heavier than a liter of air of a 1.2 gram,
will always fall back to Earth. This is due to its mass and its electrostaticomagnetic field.
After that is the buoyancy that comes into play.

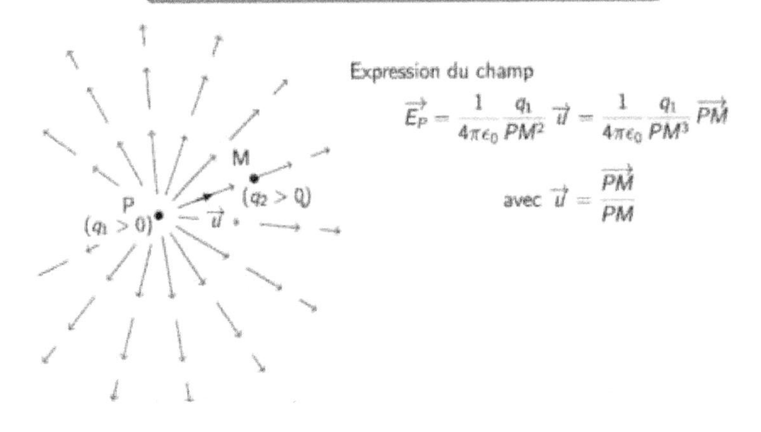

Earth is negatively charged to 500,000 electrical charges of Coulomb. The potential difference is the value of 300,000 volts (300 kV), considering the voltage between the ionosphere positively charged and the surface of the Earth. There is also a constant current of electricity, of the order of 1350 amperes (A), and the Earth's atmosphere resistance is about 220 ohms. This gives an output power of about 400 megawatts (MW), which is regenerated by solar activity.

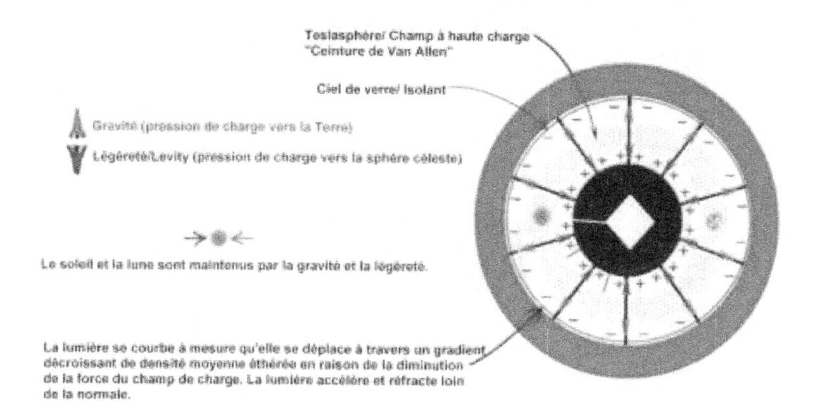

This affects the ability of the Earth's ionosphere and layers lower, causing a storm. Stored electrical energy and stored in the Earth's atmosphere is about 150 gigajoules (GJ). The system of "Earth-ionosphere Acts as a giant condenser with a capacity of 1.8 Farad.

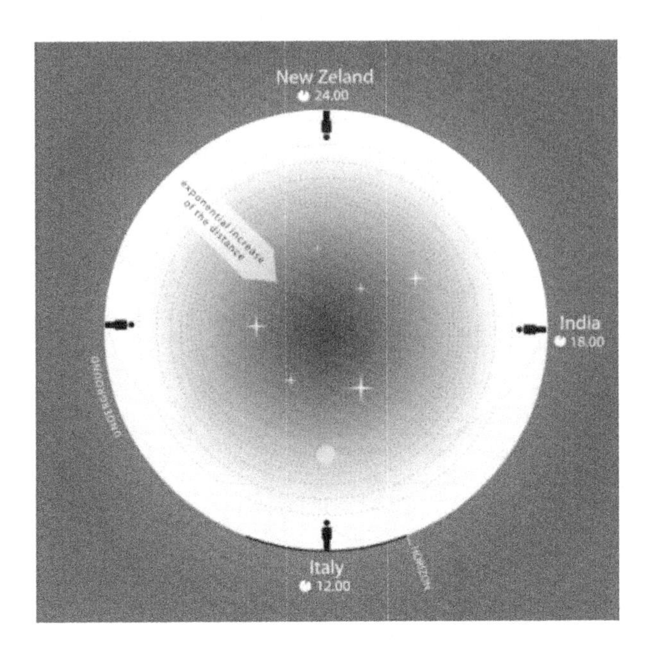

The atmospheric environment has characteristics electrostatic. Beyond an altitude of about 60 km, the ionosphere, under the effect of the ionization of its components by the solar radiation has sufficient conductivity for it to can be considered a driver. At altitude h = 60 km so we define a sphere called electrosphere positively charged on the surface with a total charge + Q corresponding to a surface charge density σ. Earth, supposed spherical radius R and center O is also likened to a driver carrying a total load - Q uniformly distributed on the surface On average, on Earth, g = 9.8 N.kg1.
The weight value depends on latitude and altitude.

TERRE			$M = 6 \times 10^{24}$ Kg d = Rayon moyen = 6 378 Km		
Altitude	**A**	**B**	**Altitude**	**A**	**B**
0	9,81	1,000	0	9,81	1,000
100 000	9,51	0,969	1 000 000	7,33	0,747
200 000	9,22	0,940	2 000 000	5,68	0,579
300 000	8,95	0,912	3 000 000	4,53	0,462
400 000	8,68	0,885	4 000 000	3,70	0,377
500 000	8,43	0,860	5 000 000	3,08	0,314
600 000	8,19	0,835	6 000 000	2,60	0,265
700 000	7,96	0,812	7 000 000	2,23	0,227
800 000	7,74	0,789	8 000 000	1,93	0,197
900 000	7,53	0,768	9 000 000	1,69	0,172
1 000 000	7,33	0,747	10 000 000	1,49	0,151
1 100 000	7,13	0,727	11 000 000	1,32	0,135
1 200 000	6,95	0,708	12 000 000	1,18	0,120
1 300 000	6,77	0,690	13 000 000	1,06	0,108
1 400 000	6,59	0,672	14 000 000	0,96	0,098
1 500 000	6,43	0,655	15 000 000	0,87	0,089
1 600 000	6,27	0,639	16 000 000	0,80	0,081
1 700 000	6,11	0,623	17 000 000	0,73	0,074
1 800 000	5,96	0,608	18 000 000	0,67	0,068
1 900 000	5,82	0,593	19 000 000	0,62	0,063
2 000 000	5,68	0,579	20 000 000	0,57	0,058

If Newton thought gravity in a hollow Earth was impossible
because he thought the Earth was turning,
it is not she who turns but the celestial sphere.

15) Polar auroras; Proof of Concave earth and poles opening

A polar aurora, called aurora borealis in the hemisphere northern and southern aurora in the southern hemisphere, is a luminous phenomenon characterized by extremely veils colored in the night sky, the green being predominant.

Phenomenon called Steve, purple luminous curve
Caused by the interaction between magnetic fields,
aurora occur mainly in regions close to magnetic poles, in an annular zone precisely called "Auroral zone", between 65 and 75 ° of latitude.
The regions most affected by this phenomenon are the Greenland, Alaska, Antarctica, northern Canada, Iceland, Norway, Sweden, Finland and the islands Shetlands in the north of the United Kingdom.

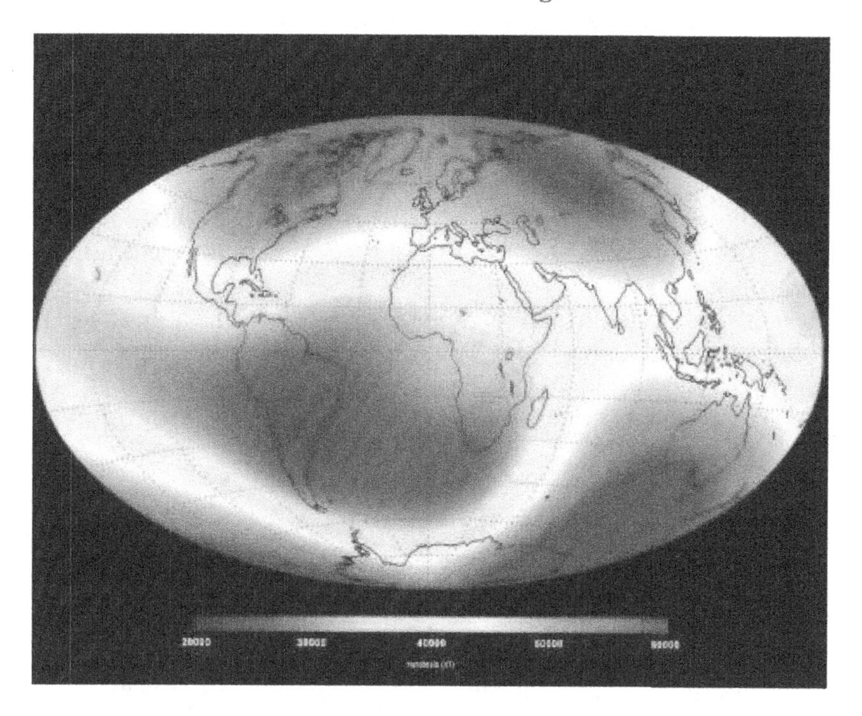

SWARM

The big problem of established science is that it does not say why this only happens in these places! Off it is obvious that if we live inside the Earth and in addition life is there possible, so it needs to be hollowed out in order to let the magnetic flux.

We can also see them all year. In the northern hemisphere we see them during the day because the sun is then on the tropic of the Capricorn, so if you want to see during the day, the period winter is better in the north. This is of course the opposite for the southern hemisphere.
The SWARM mission brought together some of unusual characteristics of the Steve phenomenon (see photo above).
As the camera was flying directly through Steve, the data of the instrument measuring the electric field have demonstrated clear changes. The temperature at 300 kilometers above the surface of the Earth has increased by 3000 ° C and the data revealed that the gas tape made 25 kilometers wide and is heading west at about 6 km / s, by ratio at a speed of about 10 m / s on the sides of the ribbon.
So the concave Earth is hollow at the north pole and the south pole, it is the reason why these observations are only possible ends of our planet.
Here is the model. Magnetic energy flows from the sphere rotating celestial and is deposited in the atmosphere, the negative energy falls to the ground (gravity), and the rest goes through the two front poles to re-penetrate the interior of the Earth.

Why planes do not fly above pole openings?

Because there are all kinds of magnetic distortions near openings because of the magnetic lines of force passing to across the inner edges of the Earth. When planes approaching the opening, their navigation instruments are going to haywire. This is due to the fact that the poles are outside the Earth.
Which would explain the adventures of Admiral Byrd who explains in his logbook, that he thinks there is a territory as big as the US on the other side of the Earth.
He had to enter the passage and found himself inside this one is outside the Earth. In the latter case, he had to

to be able to observe some time but had to go round in circles before going back inside.
What's going on outside? Are we entering a vortex?

At the moment I am more and more convinced that one lives in effect in a matrix. But a matrix created not by extraterrestrials but a creator.

About the Hollow Earth (Convex Earth)

It is absolutely impossible that the light of an outgoing sun of a hollow Earth produces polar auroras. We would observe just a light beam. These are the magnetic fields meeting who produce this phenomenon.

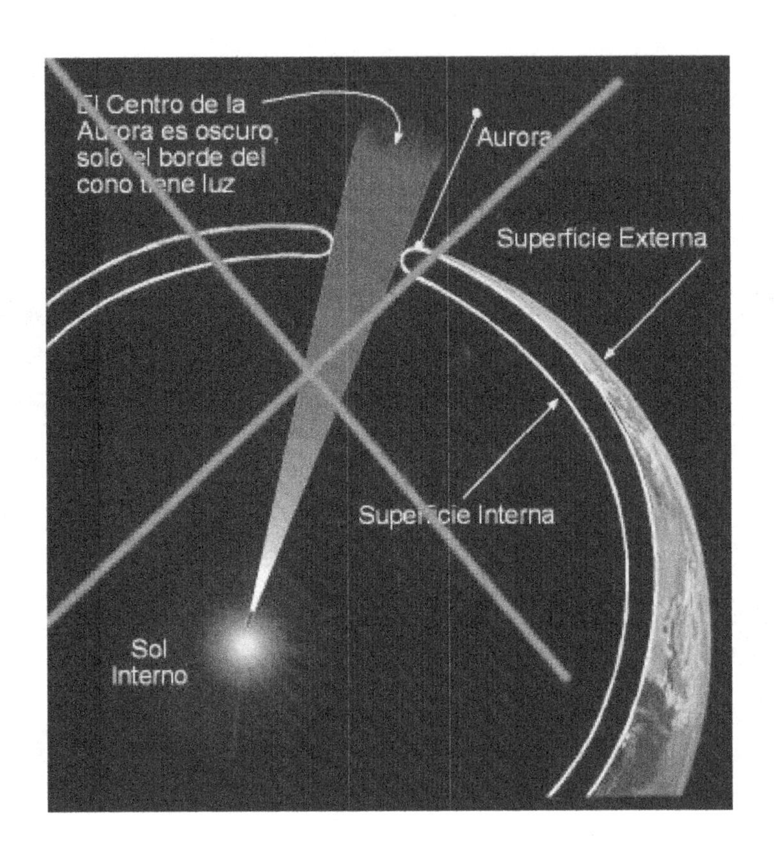
El Centro de la
Aurora es oscuro,
solo el borde del
cono tiene luz
Aurora
Superficie Externa
Superficie Interna
Sol
Interno

16) The number of gold found in the dimensions of the concave Earth

Occupancy rate of the octahedron in the sphere:
3,007 km (diameter octahedron): 4,867 km (celestial sphere) x 100 = 61.78%
Occupancy rate of the sphere in the Earth:
4,867 km (celestial sphere diameter): 12,742 km (Earth diameter) x 100 = 38.19%
61.78%: 38.19% = 1.61 (the golden ratio)

4867 Km (celestial sphere diameter): 1,618 (the golden ratio) = 3008 Km (diameter octahedron: 3,007 km)

12,742 Km (Earth Diameter): 1,618 (the golden ratio) = 7875.15 Km - 3007 Km (octahedron diameter) = 4868 Km (sphere diameter) heavenly: 4867 km)

When we divide the Earth's diameter by that of the sphere heavenly we get 2.618 which is the golden squared number!
12,742 Km (Earth Diameter): 4867 Km (celestial sphere diameter) = 2.618

4,867 Km (celestial sphere diameter): 3,007 km (diameter Octahedron) = 1.618 (the golden ratio)

3,937 km (distance Earth / Celestial Sphere) X 1,618 (the number
gold) = 6,370 km (radius of the Earth)

Digital sources

Chapter 1)
Diagram Nils Esche

Chapter 2)
Cyprustar diagram

Chapter 3)
http://www.phy.mtu.edu/alumni/history/DMGPlumbLines.pd
f
Chapter 4)
http://www.sacredtexts.com/earth/cc/cc25.htm
Earth Curve Calculator

Chapter 5)
https://fr.wikipedia.org/wiki/Courbure_terrestre
Drawing: CS

Chapter 6)
FotoMagazin n ° 11/1954
http://www.rolfkeppler.de/enewyork.htm
Bruno Carrias

Chapter 7)
https://www.cactus2000.de/fr/unit/massdrr.shtml
http://www.5thandpenn.com/GeoMaps/GMapsExamples/dist
anceComplete2.html
Stellarium
https://fr.wikipedia.org/wiki/Taille_apparente
https://sizecalc.com

Chapter 8)
http://www.innenweltkosmos.de/thema_t/t_jah.html

Chapter 9)
Schemes: Cyprustar

Chapter 10)
Stellarium
https://fr.wikipedia.org/wiki/Éclipse_lunaire
https://cyprustar.wordpress.com/2018/06/06/levraisystemes olairedelaterre

Chapter 11)
Stellarium
https://www.cactus2000.de/fr/unit/massdrr.shtml
http://www.5thandpenn.com/GeoMaps/GMapsExamples/dist anceComplete2.html
http://serge.bertorello.free.fr/optique/images/images.html

Chapter 12)
https://fr.wikipedia.org/wiki/Hélium
https://fr.wikipedia.org/wiki/Superfluidité
https://fr.wikipedia.org/wiki/Phénomène_lunaire_transitoire

Chapter 13)
https://fr.wikipedia.org/wiki/Météorite
https://fr.wikipedia.org/wiki/Soleil
http://www.wildheretic.com/whataretheastronomicalbodies
http://www.liberation.fr/sciences/2007/12/04/letrangemeteo ritedemurchison_107761
https://fr.wikipedia.org/wiki/Centrale_à_fusion_inertielle
https://fr.wikipedia.org/wiki/Disruption_(tokamak)
http://www.elementschimiques.fr/?fr/elements
https://www.sciencedirect.com/science/article/pii/S23521791 17300030

Chapter 14)
http://www.holoscience.com/wp/electricgravityinanelectricun ivecsr
http://aufzurmitte.blogspot.fr/2015/12/zellularkosmologie.ht ml
https://fr.wikipedia.org/wiki/Gravitoélectromagnétisme
http://electricityautomation.com/en/electricity/10
https://fr.wikipedia.org/wiki/Force_de_Lorentz

https://fr.wikipedia.org/wiki/Pesanteur

Chapter 16)
http://trotzderluege.foroactivo.com/t4p600dieerdeinderwirlebenundderraumderdieweltist

MORE ON FRENCH AUTHOR'S SITE
https://laterreestconcave.home.blog/

MAIL
cyprustar@gmail.com

Made in the USA
Monee, IL
25 November 2024